The Magnetosphere

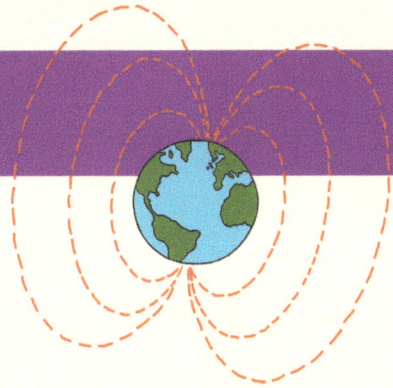

Illustrations: Janet Moneymaker
Design/Editing: Marjie Bassler

Copyright © 2022 by Rebecca Woodbury, Ph.D.

The Magnetosphere
ISBN 978-1-953542-21-2

Published by Gravitas Publications Inc.
Imprint: Real Science-4-Kids
www.gravitaspublications.com
www.realscience4kids.com

Have you ever noticed a **magnet** sticking to your refrigerator?

Why would you use a refrigerator?

For keeping cheese!

A magnet is surrounded
by a **magnetic field**.

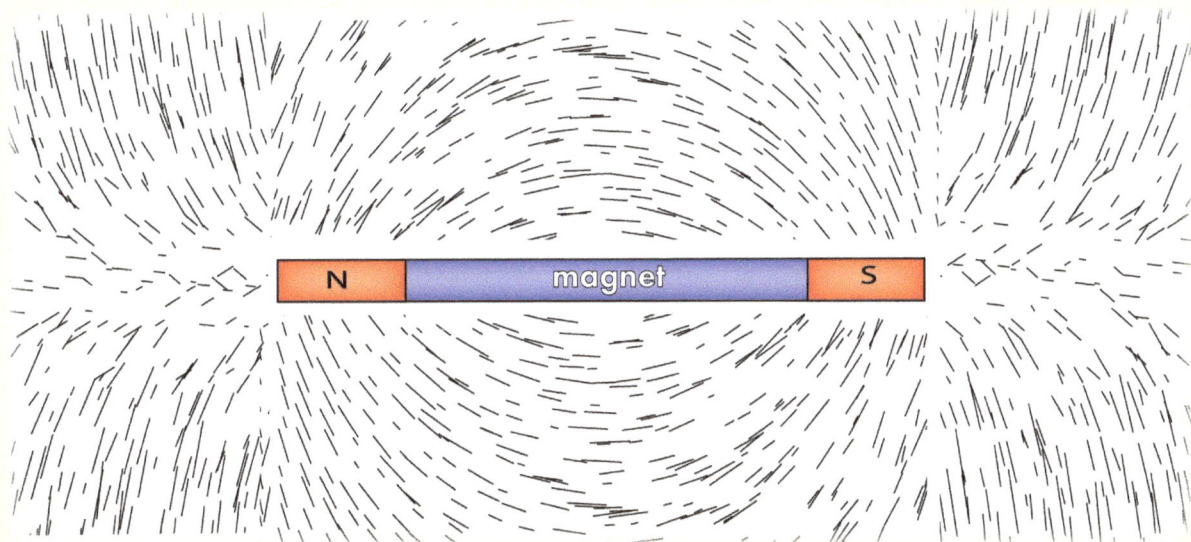

Magnetic Field

A magnetic field is the area surrounding a magnet that has **magnetic forces**.

In physics...

Force is any action that changes...

...the **location** of an object,

...the **shape** of an object,

...**how fast or how slowly** an object is moving.

(This is called the **speed** of an object.)

A magnet has two opposite ends called **poles**. Poles are created by magnetic forces going in opposite directions.

One pole is called the **north pole**. The other pole is called the **south pole**.

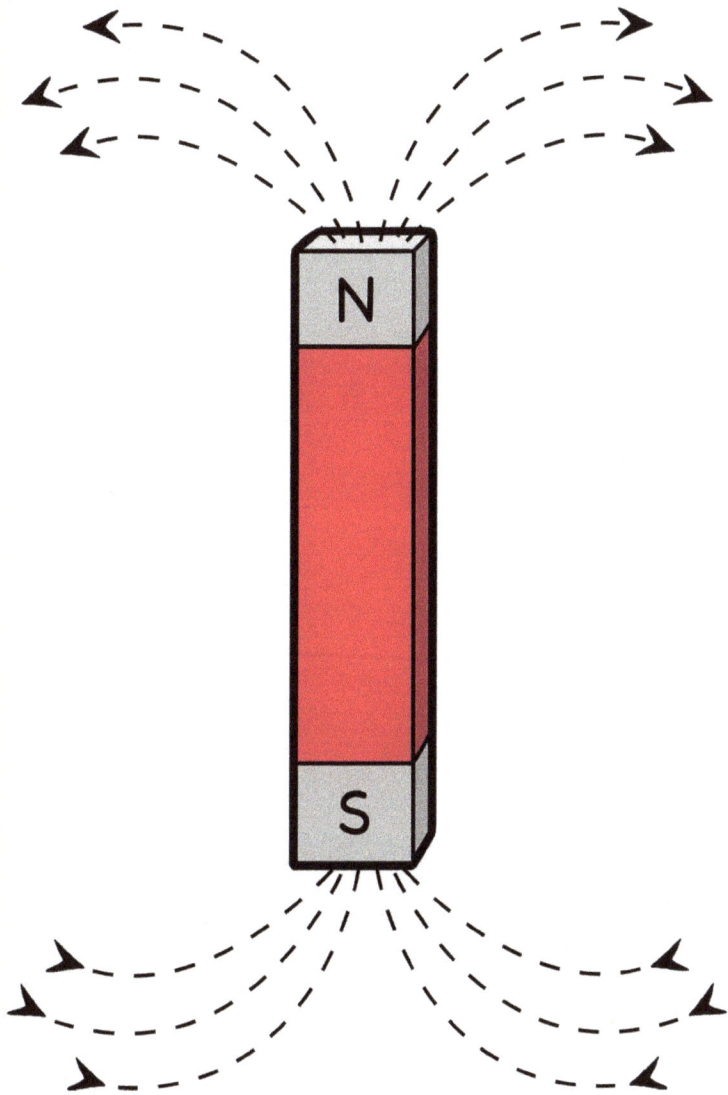

Earth is a big magnet and is surrounded by a magnetic field.

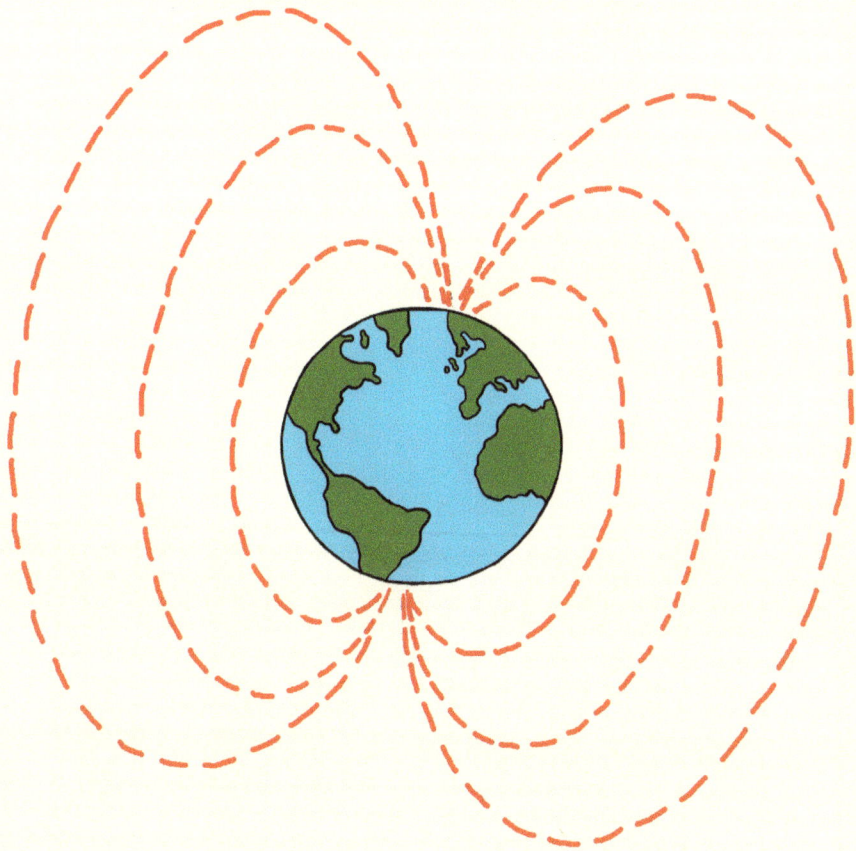

The North Pole is at the top of Earth.

The South Pole is at the bottom of Earth.

North Pole

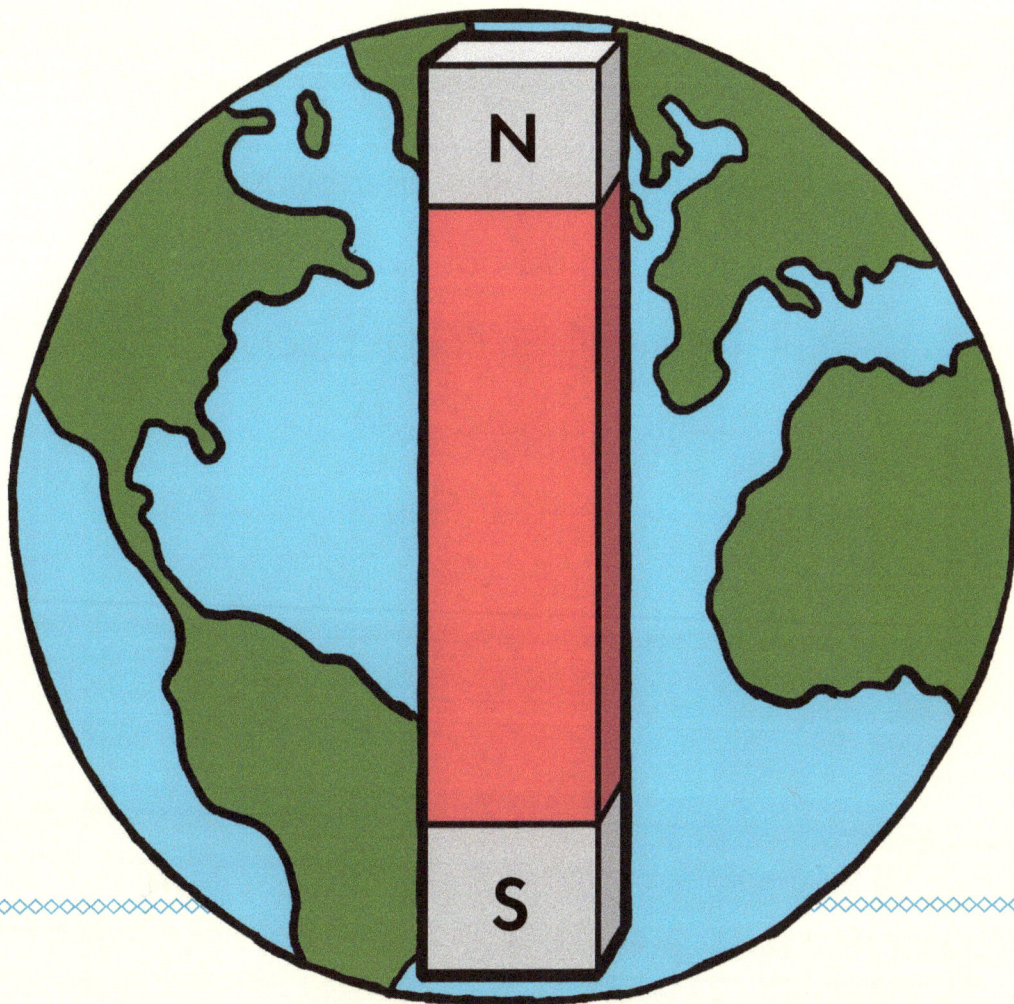

South Pole

Scientists think that the outer part of Earth's **core** is made of molten (melted) metal. The magnetic field is created when this molten metal swirls.

Wow! The magnetic field starts **inside** Earth.

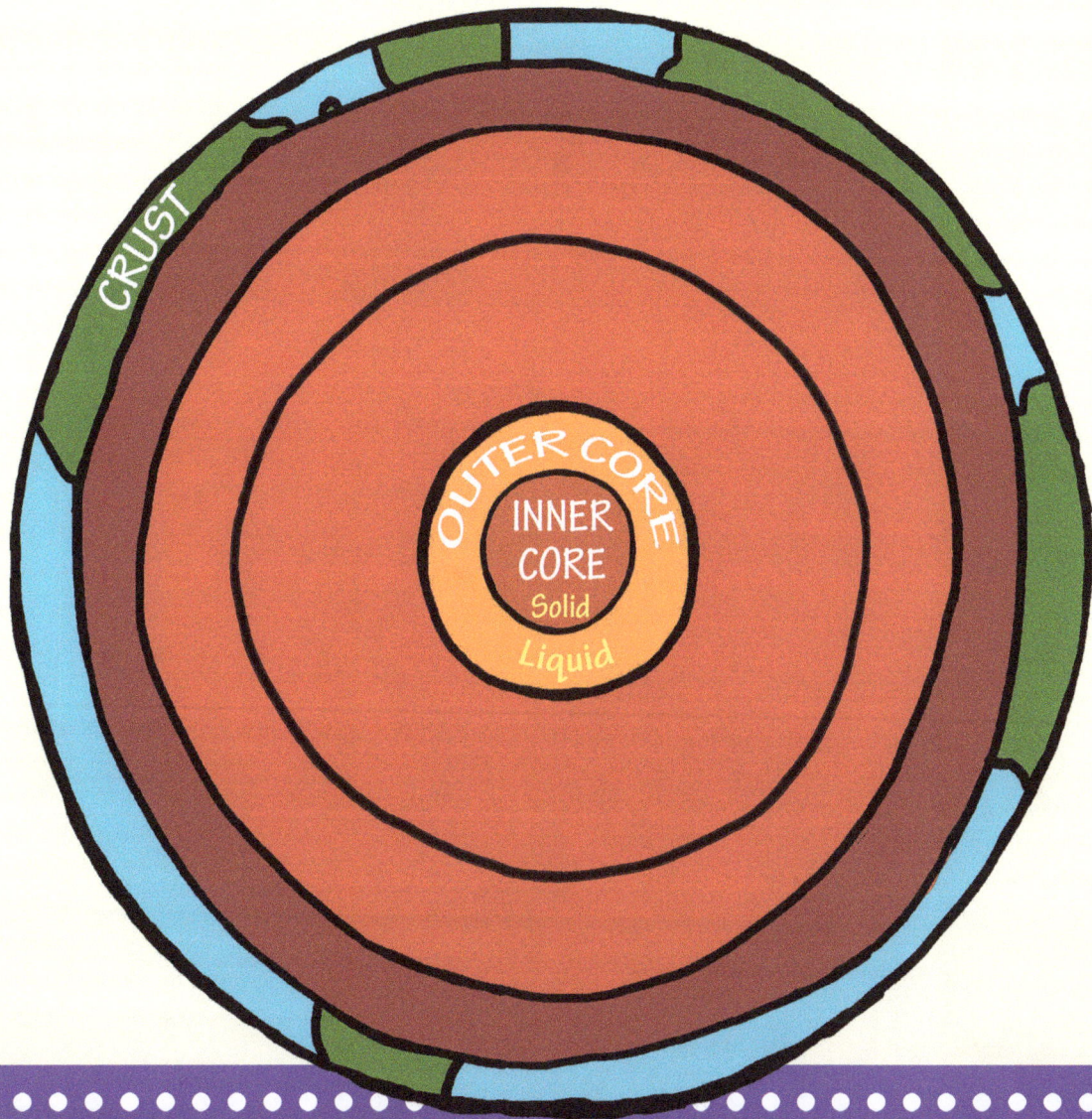

CRUST

OUTER CORE

INNER CORE

Solid

Liquid

Earth's magnetic field extends into space.

>>>

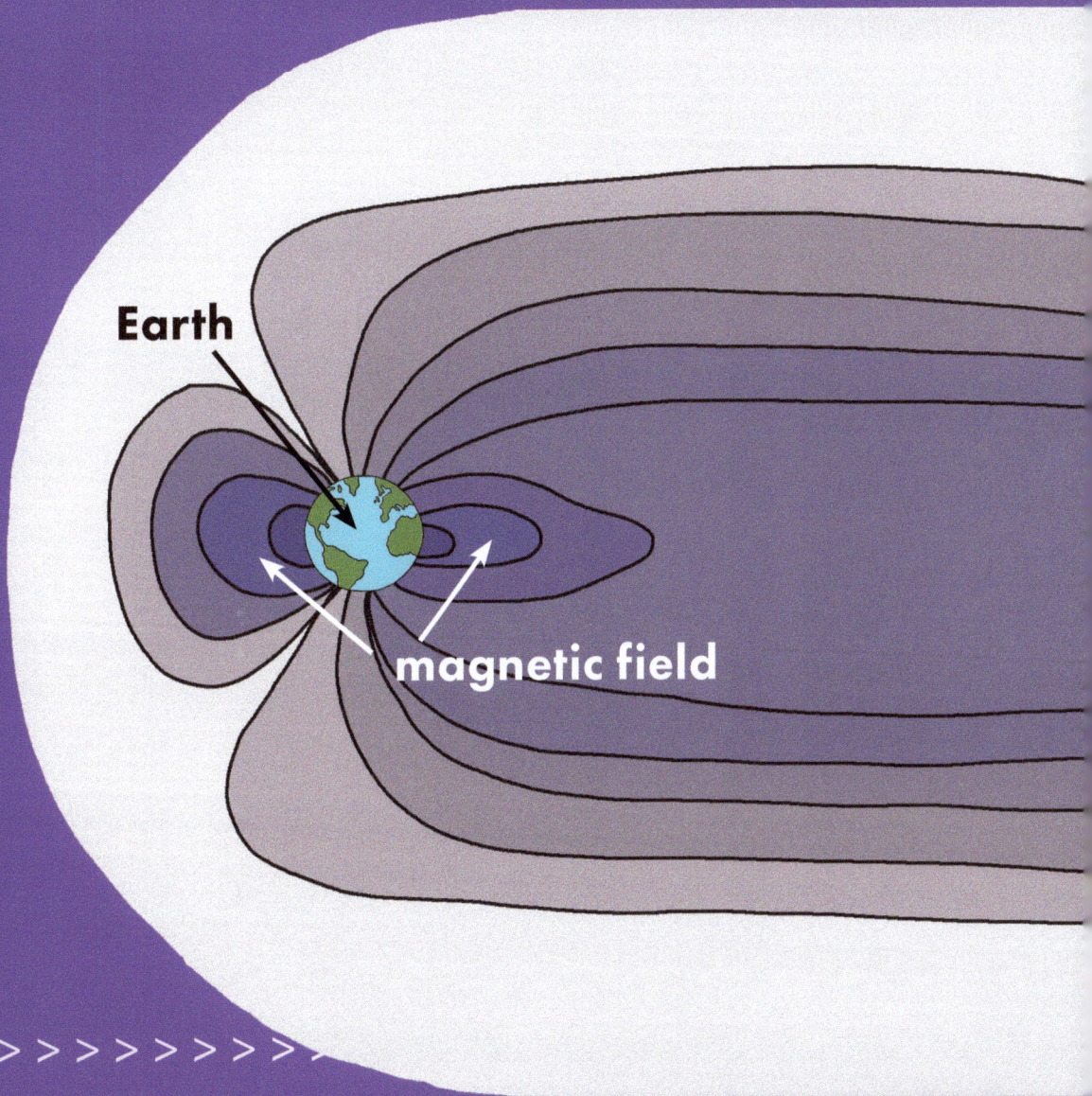

Earth

magnetic field

The **magnetosphere** is formed when **light energy** from the Sun hits Earth's magnetic field.

> >

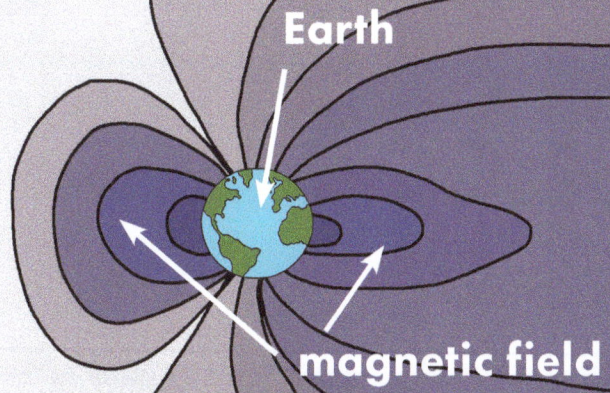

Earth

magnetic field

magnetosphere

Light
energy
from Sun

The magnetosphere is very important for life on Earth. It helps protect Earth from getting too much light energy from the Sun.

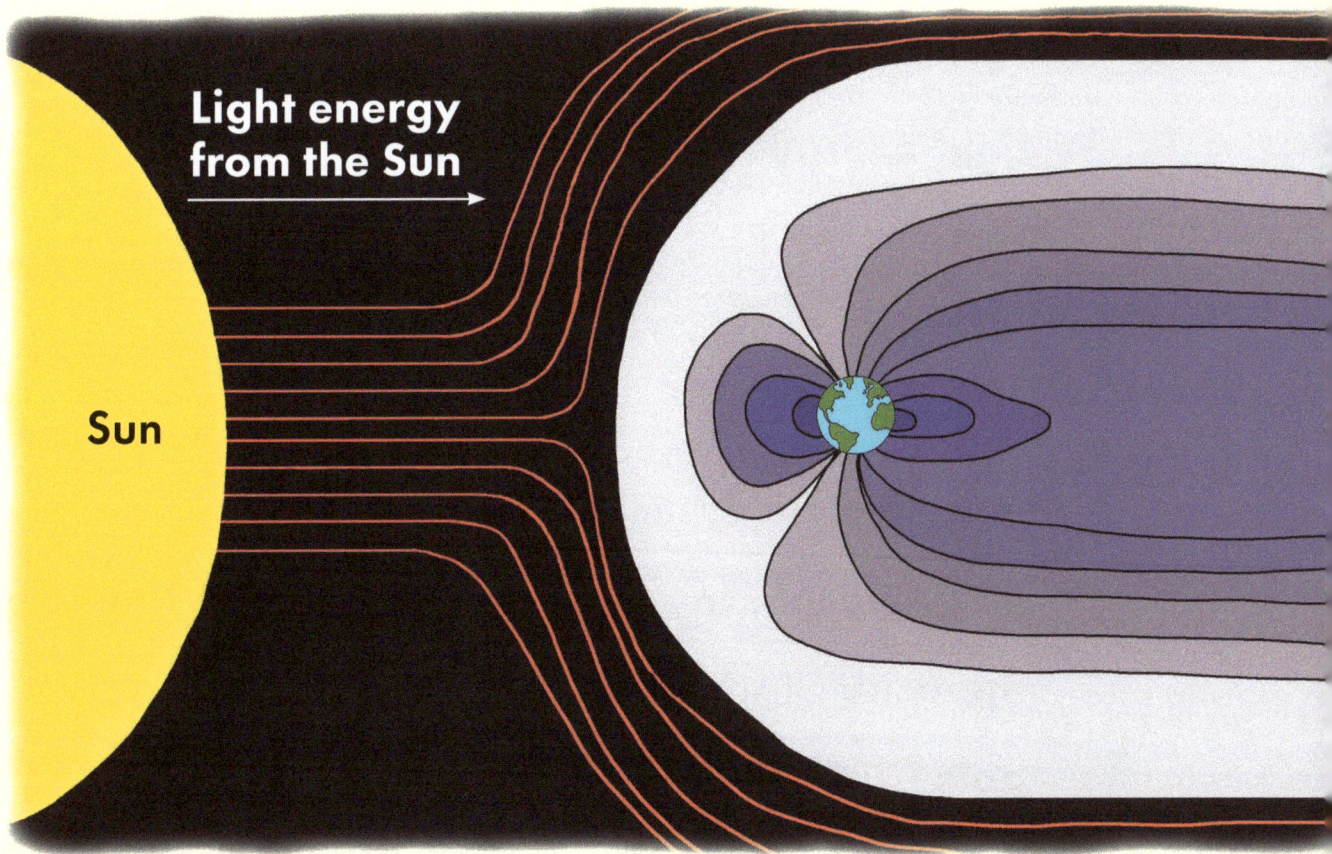

Sun

Light energy
from the Sun →

How to say science words

core (KAWR)

light energy (LIYT EN-uhr-jee)

force (FAWRSS)

location (loh-KAY-shuhn)

magnet (MAG-net)

magnetic field (mag-NE-tik FEELD)

magnetosphere (mag-NEE-tuh-sfeer)

opposite (AH-puh-zuht)

pole (POHL)

science (SIY-ens)

shape (SHAYP)

What questions do you have about THE MAGNETOSPHERE?

Learn More Real Science!

Complete science curricula
from Real Science-4-Kids

Focus On Series

Unit study for elementary and middle school levels

Chemistry
Biology
Physics
Geology
Astronomy

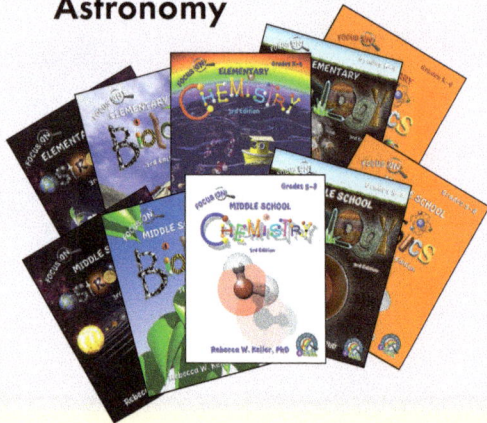

Exploring Science Series

Graded series for levels K–8. Each book contains 4 chapters of:

Chemistry
Biology
Physics
Geology
Astronomy

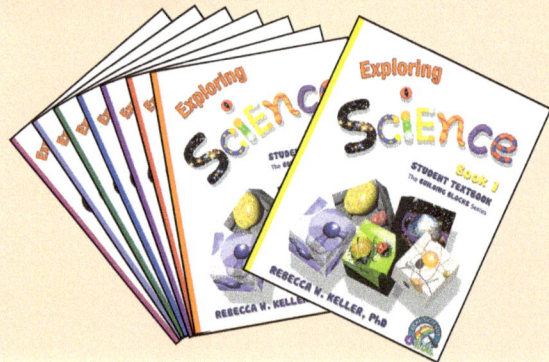